SpringerBriefs in Fire

Series Editor

James A. Milke

For further volumes:
http://www.springer.com/series/10476

Xavier Silvani

Metrology for Fire Experiments in Outdoor Conditions

 Springer

Xavier Silvani
Laboratory Sciences for the Environment
University of Corsica
Corte
France

ISSN 2193-6595   ISSN 2193-6609   (electronic)
ISBN 978-1-4614-7961-1   ISBN 978-1-4614-7962-8   (eBook)
DOI 10.1007/978-1-4614-7962-8
Springer New York Heidelberg Dordrecht London

Library of Congress Control Number: 2013939829

Printed on acid-free paper

Springer is part of Springer Science+Business Media (www.springer.com)

Contents

Symbols

d	Diameter (m)
h	Convective heat transfer (W/m^2K)
I_0	Modified Bessel's function of the first kind
k	Thermal conductivity (W/mK)
Nu	Nusselt number
r	Radius (m)
Re	Reynolds number
Pr	Prandtl number
Gr	Grassh of number
T	Temperature (°C or K)
u	Gas velocity (m/s)
ΔT	Thermocouple error (K)

Greek Symbols

β	Thermal expansion coefficient (K^{-1})
ε	Emissivity
σ	Stefan–Boltzmann constant ($5.67 \times 10^{-8}\,W/m^2K^4$)
ρ	Density (kg/m^3)
δ	Thickness of the gauge constantan foil
κ	Conductive heat transfer coefficient

Subscripts

g	Gas
GG	Gardon gauge
TC	Thermocouple
f	Film
F	Flame

Chapter 1
Introduction

Fire is a physical phenomenon in continuum mechanics involving the transport of mass, momentum, and heat in a complex medium formed by at least two phases, solid and gaseous; the fuel is usually in the solid phase and chemical reactions occur in both the solid and gaseous phases, with the flame burning in the latter. Fire creates a thermally and chemically aggressive environment with strong radiative properties. The standard diagnostics for continuum mechanics must be set up and adapted to the fire intensity. Fire research evaluates the fire hazard, ideally by predicting its consequences using numerical simulations. Within this framework, experiments have to provide a large set of data from which empirical laws can be derived. These laws will help validate numerical simulations in a wide range of parameters.

During the last two decades, a large number of laboratory-scale studies of fires spreading across fuel beds have been performed. The following references are provided as a non-exhaustive list for information purposes (Dupuy et al. 2006; Mendes-Lopes et al. 2003; Viegas 2004). Because it is difficult to design, set-up, use, and interpret results from a metrology dedicated to real-scale fires, the experimental research on the subject has used the results from fire studies at the laboratory scale. Many studies have used the cone calorimeter and the fire-propagation apparatus to derive experimental laws for fires (Kuang-Chung and Drysdale 2002; Consalvi et al. (2011); Luche et al. 2011; Bartoli et al. 2011). However, current results from recent studies (Boulet et al. 2012) also illustrate that experimental conditions using these laboratory devices deviate significantly from real fires. Thus, their results remain limited to academic considerations. These lab methods also are limited by the fire size and are not really capable of reproducing, for instance, the upwind flow effects of the wind at actual scale. In addition, they are diagnostically sensitive to fluctuations when the fire size increases. At the laboratory scale, analytical chemistry and thermal chemistry are process-engineering disciplines that also endeavour to contribute significantly to fire science: they study fire as the result of a thermal-degradation process (Leroy et al. 2010). However, this point of view presents fire physics without considering large-scale turbulent mixing of the reactive gases produced by

X. Silvani, *Metrology for Fire Experiments in Outdoor Conditions*,
SpringerBriefs in Fire, DOI: 10.1007/978-1-4614-7962-8_1,
© The Author(s) 2013

pyrolysis and their combustion. In this sense, if the measurement technologies from these disciplines are important and efficient at the laboratory scale, they must be adapted to outdoor conditions to evaluate real-scale fires. Moreover, even in this case, they cannot produce a complete overview of the fire physics, because they do not describe the hydrodynamics of the fire.

More generally, attempts to extrapolate experiments from the lab to the field scale may be doomed to failure because of the difficulty in predicting, a priori, how fire is a scale-dependent process (Pitts 1991). This is mainly due to the hydrodynamic nature of fire, which can be considered a turbulent reactive and strongly radiative flow. Inherent in this characteristic is the stochastic behaviour of governing physical quantities such as temperature, mass flow rate, and heat fluxes. These properties vary strongly in time and space. Furthermore, if fire is considered a turbulent reactive and radiative flow, its dynamics cover a range of spatial and temporal scales. This range should be extended as the size of the fire increases. Measurements of real-scale fires, therefore, are simultaneously necessary and extremely difficult to perform, because experimental devices must be able to capture both long-scale phenomena and rapid fluctuations. They consider turbulent reactive flows which are coupled strongly with other complex non-linear processes such as radiation and soot formation; fine measurements in such flows usually use optical methods to prevent flow from local perturbations. Such methods are difficult at the laboratory scale (Nathan et al. 2012). Furthermore, as we will discuss, the most important processes are often poorly understood and are blurred into a complex thermal environment. This is why it is extremely hard to develop devices that are simultaneously non-intrusive, not influenced by the fire, and sufficiently accurate in field-scale experiments.

These represent the main reasons why experiments under outdoor conditions have been neglected for a long time, despite some extraordinary and rare scenarios, such as the International Crown Fire Modeling Experiment (Butler et al. 2004), and why studies have focused instead on repeatable, short-duration, and easy-to-perform laboratory-scale experiments.

This brief will review experimental techniques for fire experiments in the field. It presents solutions, and their advantages and limitations, for studying fire in the open, beyond academic and unrealistic laboratory-scale configurations. It deals mainly with natural fires, in which fuel is formed by vegetation, but measurement techniques also can be used for confined fires or liquid fuels. As previously noted, the main problem for experimentalists working with large-scale fires is that the fire environment is thermally and chemically aggressive. However, measurement devices must be accurate because of the strong spatial and temporal variability in fire properties, especially those related to heat transfer.

Let us observe the temperature signal obtained from a wildfire experiment in southern France in spring 2006 (Silvani and Morandini 2009).

Figure 1.1 shows both the instantaneous and time-filtered curves of the temperature signal obtained with a sensor immersed in a real-scale fire of Mediterranean shrubs. The flame front spread over 3000 m² and measured 6 m high and 20 m long. The sensor was fire-resistant and therefore measurements were possible

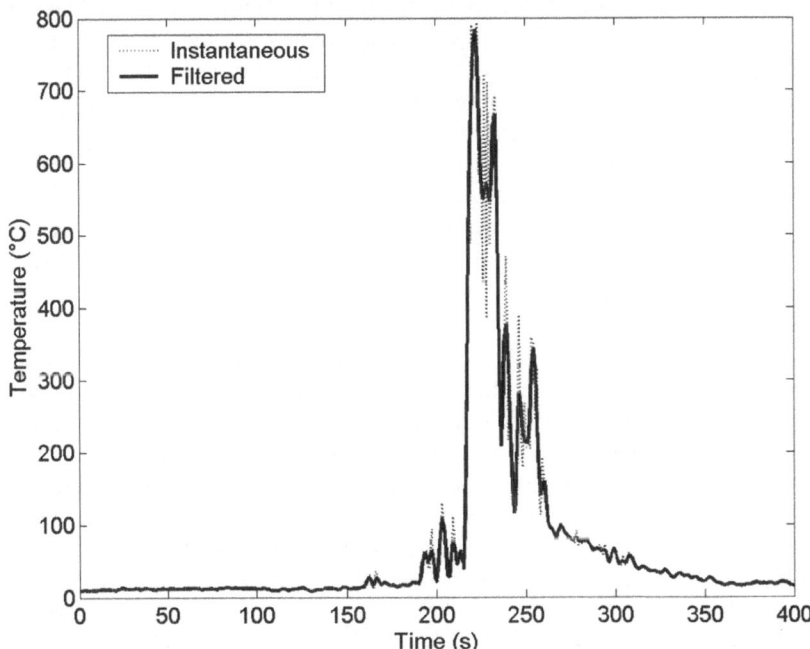

Fig. 1.1 Temperature record during a field-scale fire experiment

during the duration of the spread of the fire. One can observe that the fluctuations of this unsteady temperature signal may reach up to 400 °C around the filtered signal when its maximal value is about 800 °C. This illustrates the extent to which fire at the field scale is an extremely varying thermal flow. The wind-velocity fluctuations behave in the same manner. These fluctuations are not negligible and their contributions to 'average' quantities in fires have been discussed previously based on large-scale experiments (Silvani et al. 2009). The metrology must support a prolonged contact with the fire and be sensitive to rapidly arriving signals.

Few fire-spread experiments across vegetative fuels have been conducted to date (Carrega 2002; Viegas 2002; De Luis et al. 2004; Butler et al. 2004; Morandini et al. 2006) and with the exception of Butler et al. (2004), these experiments have provided only observations or measurements of the macroscopic characteristics of the flame front (rate of spread, flame length, flame tilt angle, and residence time, for example). In other words, these measurements have focused mainly on the geometric aspects of the flame and the kinematic of the flame front. In situ temperature and heat fluxes have been measured rarely because intrusive instrumentation dedicated to large-scale fire is difficult to develop and to operate on site. Some relevant studies have been performed (Silvani and Morandini 2006, 2009) and have produced original results describing the role of the convective heat transfer (Morandini and Silvani 2010) in fire propagation over a vegetal fuel bed. However, no empirical law can be elucidated from these different measurements because too few data have been obtained. In addition, some other fire properties remain poorly understood. For instance, the thermal radiation of the flame forming a vegetation fire has spectral properties that might not represent a black body (Boulet et al. 2009). Further investigations in this direction must be performed.

Finally, in situ measurements of large-scale fires are needed to improve and validate physics-based modelling of wildfires (Morvan and Dupuy 2004; Porterie et al. 2007; Mell et al. 2010; Linn and Cunningham 2005). Predicting fire spread with models or sub-models developed from and validated against laboratory experiments is error-prone; it should be obvious that the physical representation of the fire obtained from such experiments cannot be applied to large-scale fire scenarios. Measurements obtained from large-scale experiments are also expected to provide some guidance on fire-safety distance or wildland-urban interface dimensioning.

If measurement devices are needed for understanding the fire physics at full scale, what, then, *are* the fire physics? Fire grows and spreads by direct burning of fuels. Burning results from the impingement of a flame on gaseous and solid combustible materials and the subsequent transfer of heat towards the unburned fuel by convection or radiation. Environmental factors such as wind flow, topography, and vegetation type can affect the rate and the nature—radiative, convective or mixed—of the direction of the transport of heat. This results in fire propagation. These parameters are more or less controlled in laboratory-scale experiments (Silvani et al. 2012), but in wildfire, they are not and, as already mentioned, they can vary over a much larger range of magnitude involving drastic changes in the dynamics of fire spread. The most significant effects are observed in the

Table 1.1 Classes of fire experiments

Fire experiment	Fuel area	Indoor/outdoor conditions	Metrology
1	<10 m²	Indoor	Lab-scale
2	(10–1000 m²)	Indoor/outdoor	Lab-scale/field scale
3	(1000–10000 m²)	Outdoor	Field scale
4	>1 ha	Outdoor	Field scale

flow regime of the flame front, where the volume of reacting gases, the amount of released heat, the span of the incident radiation, and the rate of fire spread can increase by several orders of magnitude.

As a summary, Table 1.1 presents categories of fire experiments according to the area of the vegetal fuel and classifies them as laboratory- or field-scale experiments. The metrology at the laboratory scale cannot be extrapolated to every fire scenario and a new metrology must be set up established for investigating fire in real-world conditions.

These considerations have been the main motivation for this brief, which presents an overview of the most widely used experimental techniques dedicated to the investigation of fires in outdoor conditions, as well as promising new techniques. These studies are concerned mainly with fires that burn vegetal fuels but they also can be used in confined fire scenarios. We will describe several characteristics of the measurement techniques, including their design, the principles used for their instrumentation, methods used to calculate measurement errors, and the post-processing related to the devices.

The physics of fire is relevant to this framework. Therefore, we consider that the fire is a turbulent reacting flow that is exchanging mass, momentum, and energy with its surroundings. This point of view has several advantages that will be discussed below but the main one is that it includes the scale-dependant properties of the fire. The next section presents systems focusing on the measurement of heat transport. Mass flow rate and momentum are also considered, but mainly because of the role they play in certain modes of heat transfer. The third section deals with considerations that fall beyond the scope of measurements and instrumentation, namely, the experimental results derived from data processing as well as experimental methods and strategies. New sensors are presented in this third section as the context of cost management.

In the following section, we will discuss measurement systems for investigating quantities related to heat in field-scale fire experiments.

Acknowledgments The author is extremely grateful to his colleagues, Dr. Frédéric Morandini and Antoine Pieri, with whom the adventure of performing fire experiments began 10 years ago. This brief is dedicated to them. I also thank Pr. Pascal Boulet, Dr. Jean-Luc Dupuy, and Dr. Olivier Sero-Guillaume for their fruitful discussions. I also thank Arnaud Susset from the R&D vision company for helping me to understand and use optical diagnostics for reactive flows. Last but not least, a special thank you goes to Dr. Jean-François Muzy for present and future work in data processing of singular signals and the beautiful physics to which it is related.

References

Bartoli, P., Simeoni, A., Biteau, H., Torero, J. L., & Santoni, P. A. (2011). Determination of the main parameters influencing forest fuel combustion dynamics. *Fire Safety Journal, 46*, 27–33.

Boulet, P., Parent, G., Acem, Z., Rogaume, T., Fateh, T., Zaida, J., et al. (2012). Characterization of the radiative exchanges when using a cone calorimeter for the study of the plywood pyrolysis. *Fire Safety Journal, 51*, 53–60.

Boulet, P., Parent, G., Collin, A., Acem, Z., Porterie, B., Clerc, J. P., et al. (2009). Spectral emission of flames from laboratory-scale vegetation fires. *International Journal of Wildland Fire, 18*, 875–884.

Butler, B. W., Cohen, J., Latham, D. J., Schuette, R. D., Sopko, P., Shannon, K. S., et al. (2004). Measurements of radiant emissive power and temperatures in crown fires. *Canadian Journal of Forest Research, 34*, 1577–1587.

Carrega, P. (2002). Relationships between wind speed and the R.O.S. of a fire front in field conditions: An experimental example from the Landes forest. In *4th International Conference on Forest Fire Research, Gestosa.*

Consalvi, J. L., Nmira, F., Fuentes, A., Mindykowski, P. & Porterie, B. (2011). Numerical study of piloted ignition of forest fuel layer. *Proceedings of the Combustion Institute, 33*(2), 2641–2648.

de Luis, M., Baeza, M., Raventós, J., & González-Hidalgo, J. (2004). Fuel characteristics and fire behaviour in mature Mediterranean gorse shrublands. *International Journal of Wildland Fire, 13*, 79–87.

Dupuy, J.-L., Vachet, P., & Maréchal, J. (2006). Fuel bed temperature measurements in laboratory fires spreading over a slope. *Forest Ecology and Management, 234*, S117–S117.

Kuang-Chung, T., & Drysdale, D. (2002). Using cone calorimeter data for the prediction of fire hazard. *Fire Safety Journal, 37*, 697–706.

Leroy, V., Cancellieri, D., Leoni, E., & Rossi, J.-L. (2010). Kinetic study of forest fuels by TGA: Model-free kinetic approach for the prediction of phenomena. *Thermochimica Acta, 497*, 1–6.

Linn, R. R., & Cunningham, P. (2005). Numerical simulations of grass fires using a coupled atmosphere; fire model: Basic fire behaviour and dependence on wind speed. *Journal of Geophysical Research, 110*, D13107.

Luche, J., Rogaume, T., Richard, F., & Guillaume, E. (2011). Characterization of thermal properties and analysis of combustion behaviour of PMMA in a cone calorimeter. *Fire Safety Journal, 46*, 451–461.

Mell, W. E., Manzello, S. L., Maranghides, A., Butry, D., & Rehm, R. G. (2010). The wildland–urban interface fire problem—current approaches and research needs. *International Journal of Wildland Fire, 19*, 238–251.

Mendes-Lopes, J. M. C., Ventura, J. M. P. & Amaral, J. M. P. (2003). Flame characteristics, temperature, time curves, and rate of spread in fires propagating in a bed of *Pinus pinaster* needles. *International Journal of Wildland Fire, 12*, 67–84.

Morandini, F., & Silvani, X. (2010). Experimental investigation of the physical mechanisms governing the spread of wildfires. *International Journal of Wildland Fire, 19*, 570–582.

Morandini, F., Silvani, X., Rossi, L., Santoni, P.-A., Simeoni, A., Balbi, J.-H., et al. (2006). Fire spread experiment across Mediterranean shrub: Influence of wind on flame front properties. *Fire Safety Journal, 41*, 229–235.

Morvan, D., & Dupuy, J. L. (2004). Modeling the propagation of a wildfire through a Mediterranean shrub using a multiphase formulation. *Combustion and Flame, 138*, 199–210.

Nathan, G. J., Kalt, P. A. M., Alwahabi, Z. T., Dally, B. B., Medwell, P. R., & Chan, Q. N. (2012). Recent advances in the measurement of strongly radiating, turbulent reacting flows. *Progress in Energy and Combustion Science, 38*, 41–61.

Pitts, W. M. (1991). Wind effects on fires. *Progress in Energy and Combustion Science, 17*, 83–134.

Porterie, B., Consalvi, J.-L., Loraud, J.-C., Giroud, F., & Picard, C. (2007). Dynamics of wildland fires and their impact on structures. *Combustion and Flame, 149*, 314–328.

Silvani, X., & Morandini, F. (2006). Intrusive measurement system for fire experiments at field scale. *Forest Ecology and Management, 234*, S104–S104.

Silvani, X., & Morandini, F. (2009). Fire spread experiments in the field: Temperature and heat fluxes measurements. *Fire Safety Journal, 44*, 279–285.

Silvani, X., Morandini, F., & Dupuy, J.-L. (2012). Effects of slope on fire spread observed through video images and multiple-point thermal measurements. *Experimental Thermal and Fluid Science, 41*, 99–111.

Silvani, X., Morandini, F., & Muzy, J.-F. (2009). Wildfire spread experiments: Fluctuations in thermal measurements. *International Communications in Heat and Mass Transfer, 36*, 887–892.

Viegas, D. (2004). Slope and wind effects on fire propagation. *International Journal of Wildland Fire, 13*, 143–156.

Viegas DX, Cruz M. G., Ribeiro, L. M., Silva, A. J., Ollero, A., Arrue, B., Dios, R., Gómez-Rodríguez F, Merino L, Miranda A. I., & Santos, P. (2002). Fire spread experiments. In *4th International Conference on Forest Fire Research, Gestosa.*

Chapter 2
Measurement Systems

From a thermodynamic point of view, fire is an irreversible process during which vegetal cover evolves from one equilibrium state (unburned) to another (partially or totally burned). Therefore, fire deals mainly with quantities that are emitted, transported, and absorbed by material targets. The wildland fire–urban interface (WUI) (Mell et al. 2010) is concerned mainly with the heat transport from a burning forest or wildland to an urban area, including its radiation impact and firebrands (Manzello et al. 2006). Although they represent an important question of fire safety, firebrands flow as a dispersed fluid medium in which burning particulates are transported. We will see that tracking luminous particles requires special optical devices, the use of which is not trivial because flames are strongly radiative and turbulent. This is why our focus here is mainly on continuous quantities. These quantities—mass, momentum, and thermal energy—obey a conservation property. Their formal description and subsequent measurement involve two sets of thermodynamic variables. There are first so-called extensive variables, i.e. variables such as density, momentum, and heat that change in quantity with the mass or the volume of the system. They are the most important factors in fire science because they change with the size of the fire and, therefore, should be key measurements in field-scale experiments. The second type includes the 'intensive' variables that 'measure' the departure from the equilibrium state (e.g. temperature, pressure, and gas velocity). For instance, the departure from thermo-dynamical equilibrium is more intense in a Lox/H_2 laminar diffusion flame reaching up to 3000 °C than in a candle flame, where the temperature does not exceed 850 °C. In a modern thermodynamic description of the phenomenon, the fluxes of extensive quantities can be interpreted as being dependent upon the gradients of intensive variables (Jou et al. 1996). The main example for fire is the heat flux, i.e. the amount of heat by volume unit that passes through a unit surface each second, which is related to the temperature gradient (in fact, to the gradient of the inverse squared temperature, known as the affinity). Heat flux density is the extensive variable and temperature (or affinity, in more general terms) is the intensive one.

A central question in fire science is the amount of heat that is released by the fire. In laboratory-scale conditions, equipment such as the cone calorimeter and

X. Silvani, *Metrology for Fire Experiments in Outdoor Conditions*,
SpringerBriefs in Fire, DOI: 10.1007/978-1-4614-7962-8_2,
© The Author(s) 2013

the fire-propagation apparatus can offer insight. One can measure directly the total amount of heat released—the heat release rate (HRR)—by the fire as proportional to the overall oxygen consumption rate. However, such diagnostics cannot offer any information about fire in outdoor conditions where vegetal burning surfaces cover further hundreds of squared metres. Such a global quantity is less significant and hard to interpret at the field scale, whereas the local properties of the fire (local temperature and heat fluxes along the fire lines) will determine the fire hazard and guide fire-fighting processes. HRR, instead, is important for investigating the thermo-chemical properties of fuels and the fire-ignition conditions in repeatable and stable situations. At the field scale, the fire hazard is determined by the amount of heat transported towards particular targets and, therefore, fire science focuses on the heat flux. Furthermore, the fire position also is set accurately by the gas-phase temperature, which reaches up to 300 °C when gases are emitted from the thermal degradation of solid fuel and to about 800 °C in the flame. Therefore, the intensity of heat fluxes, given by temperature gradients and fire front kinematics are immediately available from multiple temperature measurement points. This is why we begin with temperature measurement systems.

2.1 Heat Measurements: The Temperature

2.1.1 Temperature Measurement

Temperature is the first quantity to measure in fire experiments at a real scale. As in radiation heat transfer, it is a dominant parameter of chemical reactions in many combustion systems, and exhibits fourth-power influence. In fire, temperature also allows the pyrolysis and flame fronts to be located; from this, the rate of fire spread can be ascertained. From the horizontal and vertical temperature distributions, the flame structure and its evolution can be observed. In addition, temperature gradients provide information about heat fluxes and temperature fluctuations reveal features of reactive flow motions (Silvani et al. 2009). In many combustion systems, measuring the temperature is a challenging problem, especially when it reaches as high as 2000 or 3000 °C. In outdoor fires, where flames are naturally ventilated by the surrounding air flow, the temperature is not expected to be greater than 1500 °C. Rather, the fire context is determined by the strong coupling of the heat radiation due to soot and/or fine-sized particles and the flow motion.

Because of the local nature of the quantity 'temperature', its measurement becomes subject to several errors due to local heat transfer and fine-scale phenomena that might be relevant in every case.

This local nature of the quantity 'temperature' has other consequences. Pointwise measurements usually are easy to perform and robust in small-scale environments. In fire scenarios, they also lead to an increase in the overall cost of the experiment because their number must increase with the fire size. Measuring a temperature field, therefore, becomes a significant issue in large-scale fire scenarios. However, this is based on modern optical diagnostics that are not really designed for use in such conditions, even if recent progress illustrates their future potential.

2.1.2 Devices for Temperature Measurement

Temperature can be measured using direct (intrusive) and optical (non-intrusive) devices.

The direct measurement devices are mainly thermocouples, i.e. electric wires of small size ranging from a few micrometres up to millimetres. Thermostats and thermometers with resistance (such as the platinum probe PT 100) are not convenient for sensing a fire at field scale.

Thermocouples are manufactured by linking two different types of metals with a joint. When the two junctions of an open circuit formed out of two different metals are at different temperatures (Fig. 2.1), a difference in potential exists between the two junctions as a linear function of the temperature difference. This is called Seebeck's effect. The coefficient S_{12}, expressing the linear dependence of voltage to temperature, is Seebeck's coefficient (Eq. 2.1).

$$S_{12} = \frac{dV_{12}}{dT_{12}} \qquad (2.1)$$

where dV_{12} (resp. dT_{12}) is the voltage (resp. temperature) variation between points 1 and 2. Seebeck's coefficient is a characteristic of the coupled metal that forms the thermocouple.

The main parameters of a thermocouple are the coupled pair of chosen metals (i.e. Seebeck's coefficient and, therefore, a temperature range), the nature of the 'hot' junction between wires, and the response time. The temperature range corresponds to the category of thermocouple. Indeed, each couple of selected metals has a Seebeck coefficient. For fire of vegetal fuels, the temperature range is usually that of the K-type thermocouple (chromel–alumel), i.e. up to 1372 °C. Some

Fig. 2.1 Principle of measurement using a thermocouple

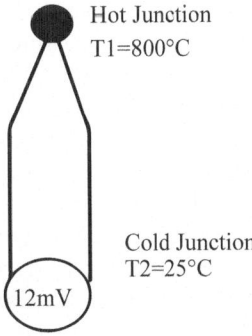

Hot Junction
T1=800°C

Cold Junction
T2=25°C

12mV

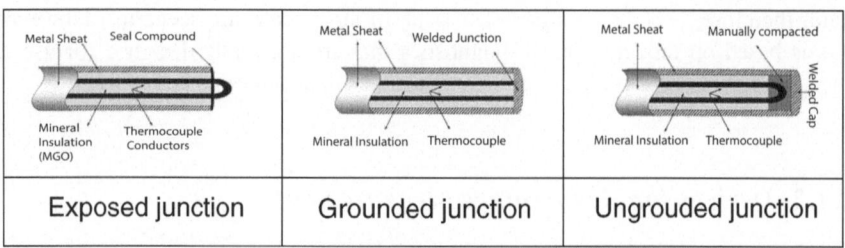

| Exposed junction | Grounded junction | Ungrouded junction |

Fig. 2.2   Three different junctions for thermocouples

cautions must be taken for their use, particularly guaranteeing the control and the independent measurement of the cold wire junction if an absolute temperature is required and not just the difference between the two junctions.

The hot junction may be exposed, ungrounded, or grounded (Fig. 2.2). An exposed junction has nothing (no protective assembly or even a tube) to cover the junction. Exposed junctions have the fastest response time, the lowest radiation error, and the least conduction error (discussed below). They suffer from corrosion and fragility. There is an additional risk in using them in outdoor fire conditions. Exposed-junction thermocouples also are prone to picking up parasitic electromagnetic signals, but the risk seems low in outdoor natural conditions. However, there are solutions to guard against this.

A grounded junction is similar to an exposed junction, except that a protective metallic sheath encloses the elements and insulation. In a grounded junction, the thermocouple wires are welded directly to the surrounding sheath material. A grounded junction is more capable of tolerating physical and mechanical abuse. It is also more resistant to corrosion and oxidation. However, thermocouples with grounded junctions suffer from a slower response time and are more sensitive to errors than are exposed-junction thermocouples. Like exposed-junction thermocouples, they may pick up stray electromagnetic signals.

These considerations—wire size and junction model—directly influence measurements made using the response time. Indeed, in a measurement system where gained voltage is converted into a digital signal for recording and monitoring, the ability to capture fast phenomena depends on the cumulative response time of the data-logging system and the response time of the connected device. Standard data loggers have a measurement speed of about 200–250 μs, i.e. are able to capture rapid phenomena at 8 kHz (according to Shannon's rule).

When a thermocouple is suddenly submitted to the temperature T_1, namely a step from the temperature of the cold junction T_2, there is a delay, due to thermal inertia, in reaching up to 63.2 % of the final value according an exponential process. This delay is the response time or the time constant of the thermocouple.

Evaluating the response time of a thermocouple is possible through the derivation of the local thermal model of the thermocouple, which is subject to all of the heat transfer in a fluid medium when the temperature changes from T_2 to T_1. Current models present the constant τ depending on the junction density ρ, the volume V, the specific heat c, the convective coefficient h, and the junction area A_s, according to Eq. 2.2 (Bentley 1998).

$$\tau = \frac{\rho V c}{h A_s} \tag{2.2}$$

The air temperature was measured using a K-type sheathed thermocouple with 50 μm wire diameter. The fine wire thermocouples have low response times and allow one to follow, with good accuracy, the rapid fluctuations of the gas temperature inside the flame when signals are sampled at 1 Hz, which is the current high frequency in outdoor environmental applications, including meteorological measurements. However, this set-up is limited when investigating high-frequency temperature fluctuations (beyond 10 Hz), which are usually neglected in fire studies at real scales and may have some relevance to models (Silvani et al. 2009). The investigation of every extensive or intensive quantity in real-scale fires is related to the fluctuating features of these fires; the time constant informs the selection of a convenient measurement tool that takes into account these fluctuating aspects. In other words, using a wrong time constant –therefore, a wrong size of the thermocouple–may filter out some fluctuations, even if these are involved in the fire.

Thermocouples are, therefore, used commonly in fire research to measure gas temperatures, but often fail to measure the true gas temperature (Cox and Chitty 1985; Luo 1997; Dupuy et al. 2003; Morandini and Silvani 2010; Silvani et al. 2012). Radiation effects that depend on the measurement conditions are considered the most significant source of errors. In fire science, discrepancies between true gas and measured temperatures are acceptable because they are less than 10 % for the probes used (Silvani and Morandini 2009). The corrected temperature curves are not provided here but radiation effects on the temperature measured using thermocouples must be kept in mind. Measurement errors with thermocouples are detailed in Sect. 2.1.3.

A good example of temperature measurements during fire experiments in the open is given in Morandini and Silvani (2010). This work consists of a series of fire experiments in the field using a vegetal fuel. Two regimes of fire spread are identified. In the buoyant regime, flame fronts are quite vertical and the fire spread is governed by the thermal radiation (Fig. 2.3). In this case, three air temperature regions are measured during the fire spread (Fig. 2.4), namely the preheating, flaming, and charring regions. The measured temperatures start at the ambient temperature and increase to a maximum of about 800–900 °C, which is an usual temperature range for burning vegetal fuels. The temperature curves show a slow trend modulated by fast fluctuations. The slow trend, namely the low-frequency part of the signal, is related to the fire spreading whereas the fluctuations are due to flame pulsations and wind gusts (Morandini et al. 2006). Details about filtering are available in (Morandini and Silvani 2010) and extensively discussed in Sect. 2.3 of this brief, 'Post-processing'.

The flame-residence times measured for each experiment are defined as the time during which the temperature is greater than 500 °C, which corresponds to the visible flame temperature. The flame residence times do not show significant differences and do not account for the two fire spread behaviours—a

Fig. 2.3 Flame fronts governed by buoyancy

radiation-driven regime and a mixed-radiation convection driven regime—that also have been identified as involved in laboratory-scale experiments about slope effects on fire (Silvani et al. 2012; Dupuy and Marechal 2010).

In the radiation-driven fire-spread regime, the air temperature remains close to the ambient value before the arrival of the fire front. The air temperature increase occurs at the time of the arrival of the fire front and the rise rate is high. The corresponding flame fronts [see the buoyant fig. flame front (Morandini and Silvani 2010) were close to vertical and the smoke plume was guided upward. It should be noted that peculiar temperature behaviour could be observed (see experiment 2; the sudden temperature drop to 200 °C for about 20 s is due to the presence of discontinuities in the fuel). In the second group of experiments, in the radiative–convective regime of fire spread, the rise rate of temperature during preheating was lower.

The increase in the air temperature occurs over a longer period before flame contact (more than 100 s prior to ignition) and temperature measurements show high fluctuations.

Another key feature of the measurement is the temperature fluctuations, which are related to the turbulent properties of the flame front as a reactive flow. The question of how to obtain an appropriate measurement system with a fast response

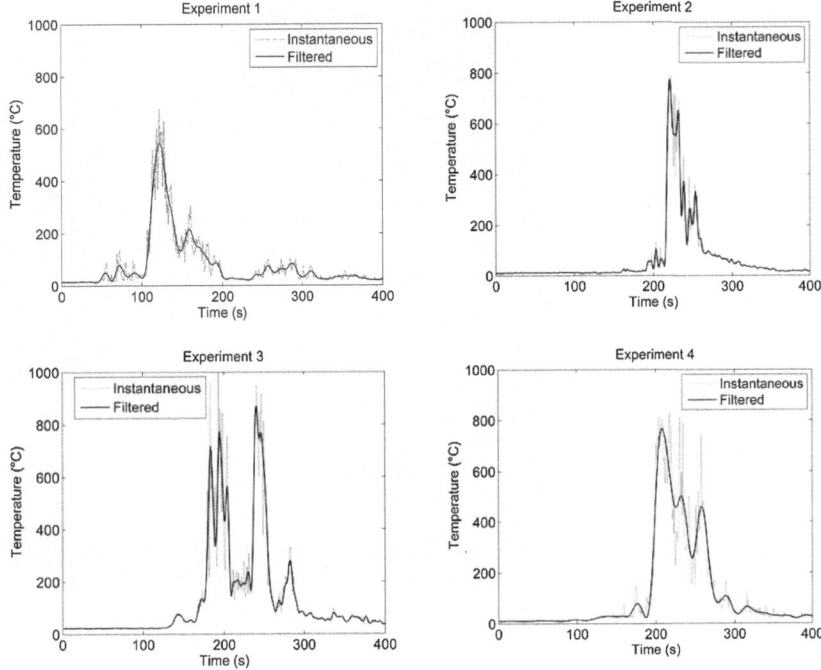

Fig. 2.4 Gas temperature in the case of previous fires (at the *top* of the vegetation)

time is addressed in the work of Frankman et al. (2010), which was pioneering research in this framework. Silvani et al. (2009) also observed that the modelling of temperature fluctuations and, therefore, their detection in fire experiments, is of primary importance for avoiding large errors in modelling the related heat flux. With regard to thermocouples, the first technological gap to discuss is reducing as much as possible the time constant to allow intrusive measurements of temperature. According to Eq. (2.2), the dependency to the ratio V/A shows that the time constant τ varies as the TC diameter. K-TCs with exposed junctions are now available with a 12 μm diameter and a time constant of about 10 μs. However, no study has reported the use of such a device in outdoor conditions. There are some applications for a protected junction in rocket engine tests, but this protection affects the thermal inertia of the tool and also degenerates the time constant.

Indirect (non-intrusive) techniques for measuring a temperature from a fire scenario in outdoor conditions are based on the optical properties of the medium. Infrared (IR) thermometry uses a map reflecting the IR emission of a hot or cold body viewed from the camera and compares this to a radiative reference of known temperature. In large-scale fires, IR digital information integrates the radiation from the entire flame volume (Fig. 2.5) and, therefore, it is hard to interpret this information as a temperature measurement. One can also cite the Rayleigh scattering thermometry, which deduces the temperature from the elastic scatter of light by molecules in a known mixture. This last method has been adapted to

Fig. 2.5 Infrared
thermometry: the plot is a
50 m² fuel bed of excelsior

clean reactive atmospheres but is not efficient for evaluating particles, as in fires.
Sensors based on the use of optical fibres use the property of light reflection and
measure the shift in wavelength that occurs in proportion to a temperature difference. We will not discuss these techniques, because they do not allow the capture of temperatures greater than 300 °C, which are commonly reached in fire
experiments.

2.1.3 Errors in Temperature Measurements

Temperature measurements with thermocouples are widely used in fire research
but the thermocouple readings are not representative of the true gas temperature
(Cox and Chitty 1985; Luo 1997, Blevins and Pitts 1999; Santoni et al. 2002;
Brohez et al. 2004). Radiation effects are considered the most significant source of
errors and can be more or less significant according to the measurement situation.
The thermocouple behaves differently ahead of or inside the flame front. When
a thermocouple is located ahead of the flame front, a higher temperature than in
the gas can be measured. This is due to the influence of radiation emitted from
the distant fire impinging on the thermocouple junction. Conversely, in the flame,
the thermocouple indicates a temperature that is lower than that of the gas because
of radiative loss from the thermocouple junction to the colder surroundings. Thus,
these errors are attributable to the temperature of the surroundings.

Correction methods for temperature measurements based on a double-thermo-couple probe (Brohez et al. 2004) or a shielded aspiration thermocouple (Blevins and Pitts 1999) are not easy to implement in field experiments. Nevertheless, the estimated error can be derived from an energy-balance equation on the thermocouple junction (Cox and Chitty 1985; Bryant et al. 2003; Incropera and DeWit 2002). Ahead of the fire front, the difference in mean temperature between the true gas and the thermocouple junction is given, to the first order, by

$$T_g - T_{TC} = \frac{\sigma \varepsilon_{TC}\left[(1 - \varepsilon_g)T_g^4 - \varepsilon_F F_{F-TC}T_F^4\right]}{h_{TC} + 4\sigma \varepsilon_{TC}T_g^3} \tag{2.3}$$

The two terms of the numerator represent the radiant heat emitted by the thermocouple junction and the radiation received by the thermocouple (mainly from the distant flame front).

Inside the fire front, this error (Luo 1997) is estimated by

$$T_g - T_{TC} = \frac{\sigma \varepsilon_{TC}(1 - \varepsilon_g)T_g^4}{h_{TC} + 4\sigma \varepsilon_{TC}T_g^3} \tag{2.4}$$

The convective heat transfer coefficient, h_T, for a thermocouple junction assumed as a cylinder can be obtained from a Nusselt number correlation (Luo 1997), and is given by

$$h_{TC} = \frac{k_g}{d_{TC}}(0.43 + 0.53\mathrm{Re}^{0.5}\mathrm{Pr}^{0.31}) \tag{2.5}$$

These corrections show that the uncertainties can also be reduced by the use of fine wire-diameter thermocouples. The finer the wires in the junction, the lower the amount of radiant heat received and released by the thermocouple, and the closer the temperature of the thermocouple will be to the gas temperature. The downside is the lower tolerance to physical and mechanical abuse, which must be considered in fire experiments at the field scale.

2.2   Heat Measurements: The Heat Flux

The main objective of experimental fire studies is to better understand the heat transport and transfer towards targets, either material or living, to preserve the integrity of the targets to thermal degradation. With the overall heat release, heat transport is central to every discussion related to fire, for scientific and engineering considerations. For the latter, the extinction of fire by water provides a framework for active research and development. In the fuel-storage industry, where fire safety is related to heat flux attenuation, how one monitors the fire scenario and manages the related extinction strategy is determined by the heat flux; this monitoring is one of the most challenging aspects in fire technology. The measurement of the heat flux of a real-scale fire remains, therefore, a central question in fire metrology.

2.2.1 Heat Flux Measurement

The heat flux measurement is generally based on the properties of solid metallic bodies that intrude into the medium where the fire occurs. These bodies, called heat fluxmeters, are the location of heat conduction from a face exposed to the fire to another face in cooled conditions. Temperatures are measured on both faces with thermoelectric sensors serving as thermocouples. The temperature gradient arising between the face heated by the fire and the cold one leads to the flux impacting the sensor; an inverse-method model for relating the outer flux to the inner one allows the measurement of the temperature gradient. This principle is theoretical and the set-up of such gauges requires an experimental procedure in which the flux measured by the sensor is compared to a calibrated one. This question is central because one generally considers as a good skill the ability of a new sensor when it returns a voltage response that is linearly dependent upon the reference heat flux. The central purpose of the calibration procedure is related to the nature of the outer flux, which the sensor is designed to measure.

Fluxmeters usually are sensitive to the total heat flux, i.e. the addition of convective, radiative, and conductive heat flux impacting the sensing face. Where fire is concerned, conduction is generally neglected if the fluxmeters observe transport only in the gas phase or in the vegetal fuel, which is a weak heat conductor. In fire experiments at the field scale, it therefore is necessary to calibrate heat fluxmeters exposed to radiative and convective heat fluxes that range over a 100 kW/m² scale; for comparison, a domestic oven has a maximal heat flux of about 5 kW/m². At that point, the question of calibration is split into two parts.

- The first concerns radiation: large-scale sources for generating an intense thermal radiation exist all over the world in laboratory or industrial environments. These sources produce a calibrated heat flux according spectral properties close to the black-body emission. This points out a condition for using radiant-heat fluxmeters in fire studies: the spectral properties of the flame emitting thermal radiation must be known or, at least, considered to be close to those of a black-body emitter. At the laboratory scale, that might not be the case for vegetation fires (Boulet et al. 2011).
- The second issue concerns heat convection. As described by Pitts et al. (2006), a group of researchers observed the response of two types of commonly used heat fluxmeters (the Gardon and Schmidt–Boelter total heat flux gauges) to pure irradiation. This was performed during an international campaign with several laboratories (a round robin). They concluded that sensitivity to heat convection must be investigated through dedicated methods. Measuring total heat flux with the Gardon gauge, using radiation-based calibration, may not

be possible in a mixed environment, such as fire. Thus, facilities are necessary to calibrate the convective contribution of the heat flux when total heat flux gauges are employed. Such facilities currently are being developed, but they are restricted to low convective heat flux levels, up to 5 kW/m² (Holmberg et al. 1999). Beyond this limitation, a central issue is the current absence of a widely used experimental set-up to calibrate the heat fluxmeter for pure convective heat flux from a turbulent flame. This is due to the large degree of freedom in a hydrodynamic transport process that occurs in complex media, such as those used for fire studies and under turbulent flow conditions. Defining the calibration conditions that are able to reproduce the fine structure of a turbulent hot flow in which the thermal convective transfer to a reference target is well-known, remains an open, and challenging, issue.

2.2.2 Devices

The thermal radiation can be measured using total or radiant heat fluxmeters (HFMs). Radiant HFMs differ from total heat fluxmeters in that they have a window fixed upon the thermosensitive electrical part of the HFM. This window stops all contact heat transfer—namely, convection and conduction—between the surrounding medium and the thermoelectrical part. This window allows only the passage of thermal radiation, and, in particular, a selected wavelength bandwidth that is composed of IR radiation. Wildfire science often employs radiant HFMs with a sapphire window, which usually allows the passage of wavelengths ranging from 0.3 to 5.5 μm. This is due to the following consideration: if the flame emits as a black body, for a flame temperature of about 800–900 °C, a significant part of the thermal radiation (more than 90 %) is emitted in the 0.3–5.5 μm range. This argument completely avoids the question of temperature fluctuations. In (Silvani et al. 2009), authors reveal the extent of fluctuations: in fire experiments at the field scale, the root mean square of a single point measurement in temperature may reach about 40 % of the average temperature value. The consequence is that the radiative properties of the flame as an emitter can cover an interval larger than 0.3–5.5 μm. The spectral properties of flame radiation in the IR raise questions related to the hydrodynamic nature of the flow. One can cite the pioneering work performed at the LEMTA (France) on such properties (Parent et al. 2010; Boulet et al. 2009).

Despite this complex environment and the large amount of technical skill required by the metrology, one can use a single measurement system to obtain robust experimental data from fire experiments in the open. The measurement of the heat flux must discriminate the nature of the flux—that is, either radiative or convective—as it must be as accurate as possible to obtain a good determination of the order of magnitude and the time scale. In the absence of a relevant tool for directly measuring the heat convection, the fire community associates the radiant HFM with the total HFM by considering that the heat conduction occurs on time scales longer than those for radiant and convective transfer (Silvani et al. 2012). In the absence of significant heat conduction (in the gas phase as well as in the solid one, if the fuel is a low heat conductor such as a vegetal fuel), the total heat flux is the sum of the radiant heat flux and the convective heat flux. As we have explained already, HFMs, both total and radiative, suffer from the lack of calibration versus a reference convective flux (more than 5 kW/m²). They are mainly calibrated facing pure irradiation. Therefore, if the two sensors are installed at the same point facing a large-scale fire and are under a pure radiant heat flux (as in the case of the buoyant driven regime in Fig. 2.6), they must measure the same signals. In this case, quantitative measurement of the heat flux is guaranteed by the limited pure irradiance calibration. The correspondence of the two signals, radiative and total, leads to a relevant measurement of the heat flux emitted by the flame—which is assumed to behave as a black body—in the irradiance condition close to the conditions at which the heat fluxmeters were calibrated.

On the contrary, if the total heat flux and radiant heat flux signals differ and the total heat flux is larger than the radiant heat flux, a convective contribution exists, as is the case when sensors are embedded into the flame. However, because of the lack of sensitivity studies of pure convection, the corresponding convective heat flux cannot be considered to be the difference between the total and radiant fluxes (Pitts et al. 2006).

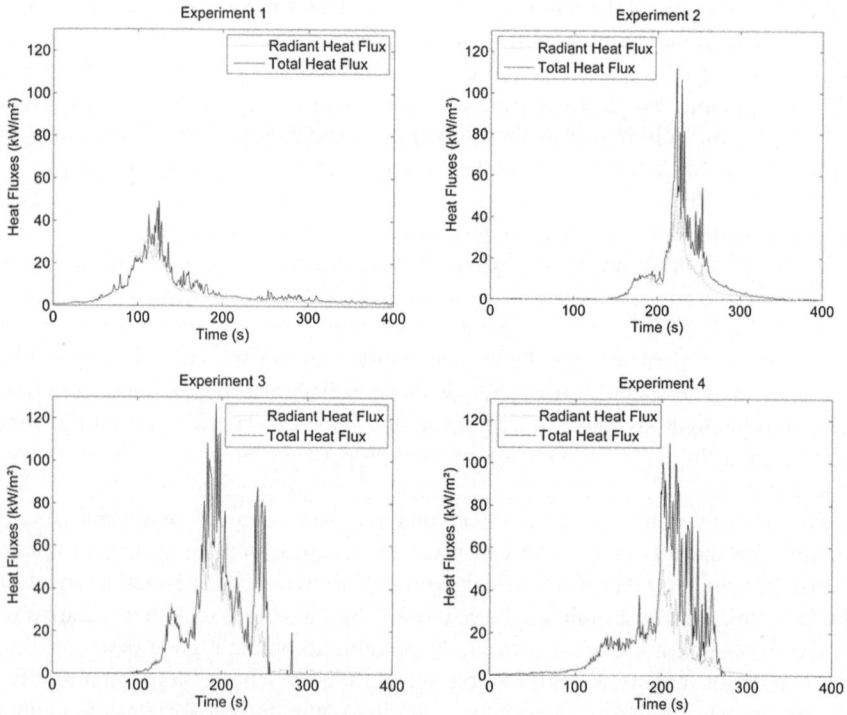

Fig. 2.6 Time evolution of heat flux densities at the *top* of the vegetation

The response of the total heat flux to strong convection, as in the case of large-scale fires, is unknown, regardless of the brand of fluxmeters that is used. A situation in which the total heat flux is lower than the radiant one signals a cooling phenomenon due to convection of the total HFM. No quantification of this cooling can be performed because of the unknown sensitivity of HFMs to convection.

If HFMs are calibrated only under pure irradiance, differential measurements of the total and radiant HFM can qualitatively detect the presence of a convective flux but they cannot quantify this characteristic.

2.2.3 Measurement Errors

The measurement uncertainties of radiant heat flux gauges with window attachments are given by the manufacturer (3 % for Medtherm Gardon gauges) when they are calibrated facing pure black-body irradiance. However, caution should be exercised when interpreting the measurement of total heat flux in a mixed radiant–convective environment obtained from the output of Gardon gauges using radiation-based calibration, because measurement uncertainties occur. In their pioneering study, Kuo and Kulkarni (Kuo and Kulkarni 1991) quantified this error, which leads to heat flux underestimation, and proposed a correction when calibration is based only on radiative heat flux. The correction to be applied is obtained from the thermal diffusion equation for the foil and can be expressed in terms of the modified Bessel's function of the first kind. This correction depends on the magnitude of the convective heat transfer coefficient. The ratio of the incident heat flux in a mixed environment and the heat flux indicated by the gauge is given by (Kuo and Kulkarni 1991)

$$\frac{q_{mix}}{q_{rad}} = \frac{(\frac{mr}{2})^2 I_0 mr}{I_0 mr - 1} \qquad (2.7)$$

where m stands for the squared root of the Nusselt number, Nu; r is the radius of the gauge; and I_0 is the modified Bessel's function of the first kind,

$$m = \sqrt{\frac{h}{\delta \kappa}} = \sqrt{Nu} \qquad (2.8)$$

This means that the ratio of the total and pure radiant heat flux measured by total and radiative HFMs is a non-linear function of the heat transfer coefficient. In the absence of any calibration in the convective conditions, this non-linear dependence is unknown.

The use of two different gauge types (total and radiant) allows for a quantitative determination of the radiant heat flux when convective heat transfers are negligible. In this case, the heat fluxes measured with total and radiant gauges are equal and no correction to the total heat flux is necessary, because the gauges are used in a thermal environment that is similar to the one in which they were calibrated. Conversely, when the heat fluxes exhibit significant discrepancies, the role of convective heat transfer cannot be neglected and the uncertainties need to be evaluated. This coefficient for a gauge foil assumed as a flat plate can be obtained from the Nusselt number ($Nu = m^2$) correlations (Drysdale 1985; Incropera and DeWit 2002). It should be recalled that the Nusselt number, Nu, is the non-dimensional form of the convective heat transfer coefficient, h. This coefficient depends on the flow properties around the gauge and on its geometry. If gas-transport properties can be evaluated from the measurement of the temperature (Bird et al. 2006), the regime and the nature of the convective flow viewed from the sensor are unknown ahead of and into the flame front. Identifying these features is not straightforward when the flame front spreads under wind-blown conditions and direct information about the velocity field is unavailable, because the gas velocity was not measured in the present study.

Several Nusselt number correlations were, therefore, tested in (Silvani and Morandini 2009) assuming a natural or forced convection regime around the HFM (Drysdale 1985; Incropera and DeWit 2002; Baukal and Gebhart 1996). From the range of Reynolds and Grashof numbers estimated using the gauge diameter, the flow can be considered to be laminar at the gauge surface, even if the flames are strongly turbulent. From the set of experiments in (Silvani and Morandini 2009), the error for evaluating the thermal radiation associated with the use of a total Gardon gauge in a mixed radiative convective environment may exceed 50 %. In conclusion, the total HFM cannot be considered to be reliable for measuring the heat flux due to both radiation and convection without a strong effort in calibrating this versus a convective heat flux.

This analysis exhibits the challenging problems that may be resolved by experimental fire science. One can summarise previous considerations by observing that two main unknowns remain for upgrading the measurement of heat transfer using fluxes: the velocity field, which can set the value of Nusselt numbers, and the gas density. The latter is a very important quantity; its behaviour signals the nature of fire in a combustion systems. In a fire, flammable gas result from the pyrolysis of solid materials (vegetal or not) and a mass flow from the solid to gaseous phases where the flame exists is set up at each instant by the 'fire system'. Contrary to burners, engines, and other 'human-designed combustion systems', in a fire at the field scale, this mass flow rate is not a steady quantity causing the fire to be a strongly turbulent flow. The mass transfer is usually called the 'source term' in fire modelling. Its rate corresponds to the kinetic of solid material degradation and involves some considerations in thermochemistry and analytical chemistry. However, neither heat fluxes nor the vegetal samples involved in these laboratory-scale analyses can be compared to those encountered in real conditions. The results they provide are limited to academic considerations.

Measuring these quantities in outdoor conditions—the velocity field in the gaseous phase and the mass transfer between solid to gas—is the greatest challenge in experimental fire sciences. Such questions relate to the ability to investigate the mass and the momentum budgets of a fire at a local scale to improve the understanding of the heat (energy) transfer.

2.3 Measuring Heat Convection and Source Term: Outlooks and Prospects

The measurement of the momentum requires measuring both the velocity and gas density in the gas phase. This measurement, if locally performed, allows for a direct measurement of the convective heat flux. Indeed, the local density of momentum is an extensive quantity defined by $\rho\,\boldsymbol{u}$, where ρ stands for the gas density and u the 3D velocity vector at the measurement point. By assuming that the heat density per mass unit is given by the product c_pT, i.e. the specific heat c_p at constant pressure and the gas temperature T, the convective heat flux is the vector field $\rho\,c_pT\,\boldsymbol{u}$. The measurement of the convection involves, therefore, the measurement of the velocity field, which is one of the most challenging questions in experimental fire science and involves the heat transported by convection. Determining the heat transfer toward a solid or a liquid fuel requires information about the contact between the flow and the fuel. It is important to distinguish between transport and transfer.

2.3.1 Measuring the Velocity Field

Fire is a scale-dependant process (Pitts 1991), with the range of the length and time scale from a chimney to a forest fire increasing by several orders of magnitude. This means that the experimental tools designed to measure thermodynamic quantities related to fire must be fire-resistant and able to capture these increasing scale ranges. In terms of the velocity field, another difficulty arises as one scales from a candle to a forest fire, because flow regime changes from laminar to turbulent both because of the extension of the scale range and the stochastic nature of the data. The velocity field shaping a flame in a real-scale fire is turbulent. Its measurement requires fast-response tools and, when possible, the ability to be non-intrusive, so that the local flow perturbation caused by contact with a solid instrument can be limited.

Pressure probes seem to be a robust measurement technique for investigating the velocity in a fire, although only a few fire studies have examined these instruments. These probes include the Pitot tube (Fig. 2.7). The velocity measurement principle proceeds from Bernouilli's theorem applied to the tube: the Pitot tube measures a velocity difference between a null velocity point (static pressure point) and a non-null velocity point (dynamic pressure point). First, however, the density must be constant over the entire tube, or corrected as a temperature function. This once again illustrates the strong coupling of gas density and velocity fields in a variable-density flow and the strong need to measure gas density. Performing the latter measurement in a fire is not a trivial issue. Second, the streamlined flow that is stopped on the static pressure point must be aligned with the tube; if this is not the case, angle error is generated

Fig. 2.7   Pitot tube

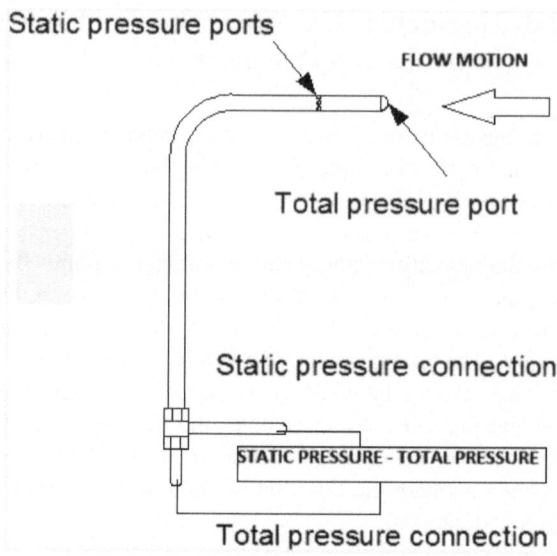

Static pressure ports

FLOW MOTION

Total pressure port

Static pressure connection

STATIC PRESSURE - TOTAL PRESSURE

Total pressure connection

that is quite impossible to correct. Finally, there is strong inertia in the pressure measurement because it requires time to transport the pressure difference that is observed all along the tube to an analogue–digital converter that provides numerical data, without leading to pressure loss along the line. Finally, the response time of the Pitot tube is linearly dependent on the dynamic viscosity and the square of tube length from the pressure point to the data logger. In a fire, the dynamic viscosity increases with the 3/2-power of temperature and, at the field scale, the hydraulic line from the Pitot tube to the data converter can be long. In practical experiments, we have measured response time to be about 0.1 s, i.e. close to that of the 250 μm TC and the radiant heat fluxmeter that we used. However, the metrology quickly becomes intrusive and the one-point unidirectional measurement that can be obtained with this technique is subject to caution. One can also find double pressure probes, such as those used by Bryant in the full-scale fire enclosure (Bryant 2009). These probes work on the same principle as a Pitot tube but, when the probe is attached to a bi-directional pressure transducer, the flow can be sensed in either direction.

New optical diagnostics such as particle-image velocimetry (PIV) now can be adapted to fire experiments. In its standard configuration, PIV involves lighting a flow containing particles to produce two sets of images separated by an inter-image interval of two laser pulses. By computing the most probable displacement of particles from one image to the next using an intercorrelation procedure, one can measure the 2D velocity field in the light plane (Fig. 2.8). In fire experiments at the laboratory scale, one can cite the pioneering contribution of Lozano et al. (2010), who found that, by seeding a fire spreading over a horizontal bed of excelsior with alumina oxide particles (which are fire resistant), strongly resolved velocity fields could be produced to investigate the vertical structure of

Fig. 2.8 Principle of
particle-image velocimetry
PIV measurement

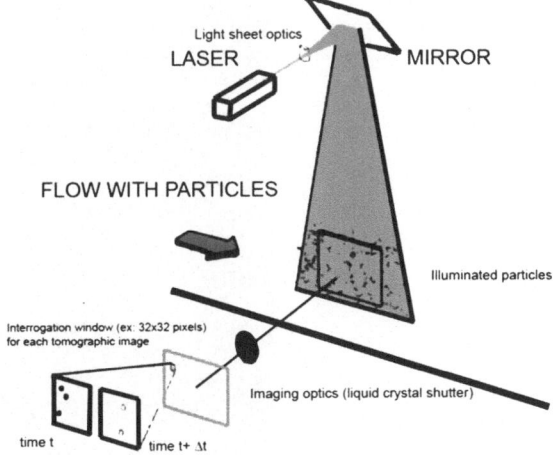

the corresponding flame. Such a technique also is feasible in outdoor conditions if special precautions are made (Fig. 2.8) (Morandini et al. 2012b).

Whatever the size of the experiment, some attention must be made to the following points. First, one has to control the flame illumination in the second PIV image (as represented on Fig. 2.9). Indeed, soot in the flame causes excessive illumination of the second PIV image, which is usually produced over a longer aperture time than the first (for reasons related to the CMOS flash). We previously proposed the use of a crystal liquid shutter with a C mount to reduce the long aperture time of the PIV camera during the capture of the second image. The quality of the second image obtained using such a technique can be observed in Fig. 2.10.

Second, to use PIV in fire experiments at the field scale, a technical solution must be found when seeding in outdoor conditions, to avoid the natural dispersion of the seeds by wind, which takes them far from the light plane. Figure 2.11

Fig. 2.9 Large-scale
particle-image velocimetry
of a fire spreading over a
vegetal fuel bed in outdoor
conditions

Fig. 2.10 Tomographic
images from pulsed-laser
lightning of a spreading flame
front

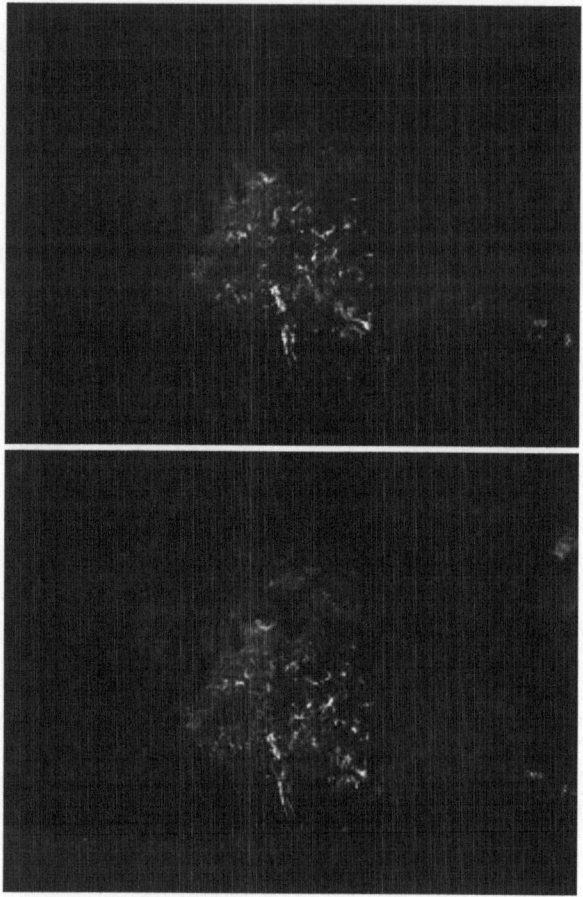

shows the displacement of the intercorrelation peak computed with the two previous images. This causes the velocity field to be measured in pixels. Scaling using
the delay between two images produces a velocity field in dimensional units.
Despite its promising features, PIV for fire experiments has some limitations. The
main limitation relates to the resolution of PIV cameras. Indeed, when all technical problems are solved (i.e. the set-up of the light plane and related optics and
the seeding system), one can produce a pair of PIV images. However, the ability
to use PIV to capture the whole range of flow scales is limited by the resolution
of the smaller flow scales possible with its cameras. Usually, for combustion, PIV
cameras have resolutions of about 2–4 Mp in each direction. However, the flow
field is not resolved at the pixel size. About 32 or 64 pixels are needed to form
the interrogation window in which the group velocity of particles is computed as
the flow velocity. In large-scale PIV, the ratio between the larger and the smaller
resolved scales is given by the ratio between the PIV window dimension and the
interrogation window dimension (Fig. 2.8). This approaches 2000/64–30 if the

Fig. 2. 11 Velocity
field resulting from
intercorrelation between
previous tomographic images
(vectors are in displacement
units)

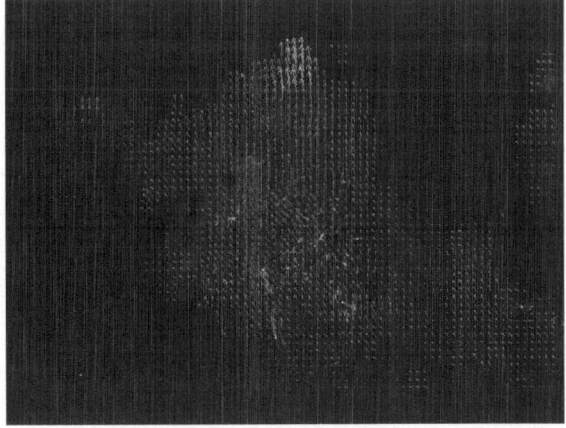

camera has a resolution of about 2 Mp per direction and if it requires a standard
window size of 64 pixels. If the PIV window is 2 m large, the flow velocity field is
resolved at 6 cm, i.e. far above the size of smaller vortices.

2.3.2   Measuring Scalars: Mass Loss Rate, Density Field, and Reaction Rate

The mass loss rate is a key parameter of fire sciences: in modelling approaches
based on empirical models, the heat release is proportional to the mass loss rate,
considering the reaction enthalpy for a mass unit of fuel. In physics-based models,
this quantity expresses the mass flux between the fuel and the gas phase. It is usually set by the temperature of the fuel under thermal degradation. At the laboratory
scale, many results are available for this measurement for fire materials, vegetal
or otherwise, using a cone calorimeter or thermogravimetric analysis (TGA). This
measurement also may be possible at the field scale using load cells that are protected from the fire using thermal insulation, although the technology is still under
development. The correlation of the measurement usually is set with the temperature, but for large-scale fire experiments, it would be more convenient if it were
correlated with the incident heat flux. The heat flux is a strongly scale-dependant
quantity whereas the temperature is not.

The measurements of scalar quantities related to the reactive mixing—that is
density, mass or mixture fraction, and reaction rate—are not trivial to perform
in the framework of turbulent combustion, for internal and enclosed geometries
(engines, burners). Some modern optical diagnostics exist, even for scenarios
with large-scale radiative flames (Nathan et al. 2012). However, as of the end of
2012, there is no evidence in the related literature that such diagnostics are likely

to be available for fire experiments in the open. The question of the density is central: how can the gas mixture—specifically, the composition and density—be measured in a fire? This is important for many reasons, including the ability to understand the gas-mixture dynamics and measuring the convective heat flux. The Fourier transform IR (FTIR) spectrometer can be used for fire scenarios; indeed, continuous emission monitoring (CEM) based on the FTIR spectrometer exists for combustion systems and provides continuous information about carbon monoxide, oxygen, and carbon dioxide in flue gases of engines and industrial burners. Because these species are expected to play a significant role in fire chemistry, CEM may be performed in fire sciences, even at the field scale, for tracking H_2O, CO_2, CO, N_2O, NO, NO_2, SO_2, HCl, NH_3, CH_4, C_2H_6, C_3H_8 and C_2H_4, C_5H_{12}, and C_6H_{14}, which have been identified as being generated by the thermal degradation of vegetal fuels. Progress currently is being made in this direction.

One can also expect the development of optical diagnostics for investigating the scalar field of a large-scale reacting flame. However, obtaining results from large-scale fire experimental scenarios is not trivial. In (Schulz and Sick 2005), light-induced fluorescence (LIF) is presented as the most robust technique for quantitative measurements of scalars including temperature in flames. First, however, because of the principle of LIF—which involves tracking by imaging the natural relaxation of molecules excited by a strong laser light—the technique requires powerful lasers that are not easy to display in outdoor conditions. Furthermore, the investigation windows, which cannot exceed tens of centimetres in size, are rather hard to select in a flame front that is several metres high. Furthermore, with this lighting technique it is rather difficult to eliminate interference from particles larger than molecules.

Despite these difficulties, determining the fraction of given reactive species in the turbulent motion of the flame remains a good tracer of the mixing dynamics. In this framework, we recently performed small-scale experiments of a fire spreading over a fuel bed of excelsior to test the coupling of optical diagnostics (Morandini et al. 2012a). At the time of the writing of this brief, there was no reference in English. In (Morandini et al. 2012a), using an optical diagnostic to investigate the chemical field of a spreading fire was a new technique; despite its limitation to a laboratory scale, we present the main results here because of its promise.

The technique, called chemiluminescence OH*, consists of capturing the natural emission of the OH* radical during the combustion process using an intensified digital camera (ICCD) with filters. Because of the intense radiation of a fire flame over a large wavelength band (mainly in the IR), the spectroscopy of a fine ray in the UV range is technically hard to obtain. However, this technique, when coupled with an optical diagnostic for flow measurement such as PIV, may produce precious information about the mixing process during combustion and the influence of external parameters such as slope and wind flow. Figure 2.12 shows why the UV radiation of OH* radical is a good tracer of the reaction zone.

Fig. 2.12 Spectral properties of radiation for flame spectroscopy of a fire spreading over 1 m² excelsior

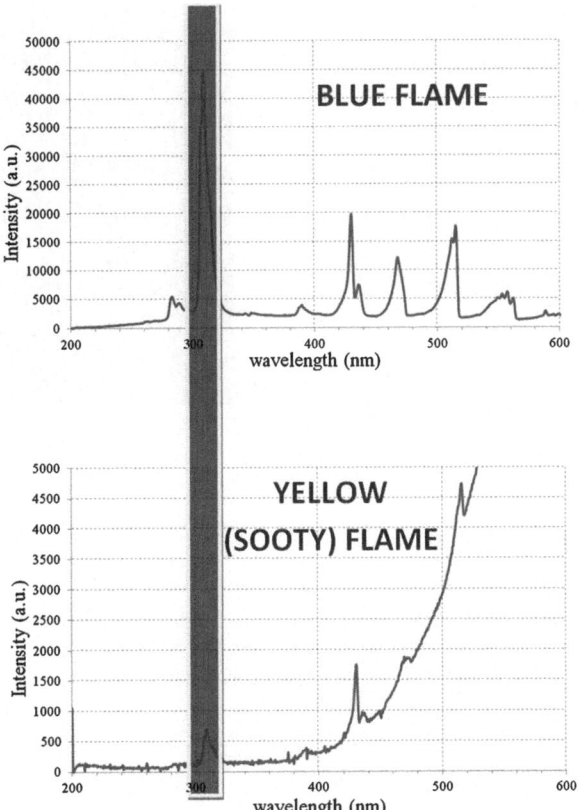

The emission occurs in the UV range, which means that it is sufficiently far away from the contribution of the thermal radiation in sooty flames. The modified camera is protected with a bandwidth filter centred on the UV peak of OH* emission.

In the field of turbulent combustion, some geometric configurations indicate that the reaction rate is linearly dependant on the OH* intensity. This is the case when the flame is stabilised in a conical shape over a constant mass flow rate, as with a burner. However, this can no longer be assumed in an extended fire front because the ICCD diagnostic for OH* chemiluminescence is an integrated diagnostic. From this point of view, the technique suffers from the same limitations as IR imagery for temperature measurement. In this sense, it is important to recall that the OH* field represented on Fig. 2.13 is integrated in the flame volume and not captured the intercept with a light plane.

Despite the difficulties inherent in using these optical diagnostic techniques to study fire, they may represent the future in fire science, allowing us to explore the internal dynamics of the mixing and combustion processes in the gas phase.

Fig. 2.13 Intensity of
OH* emission captured by
chemiluminescence

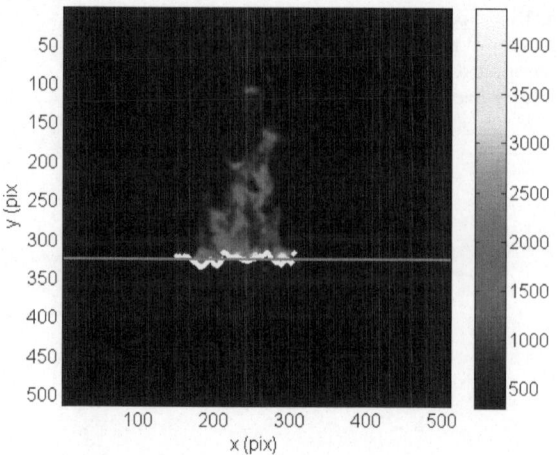

References

Baukal, C. E., & Gebhart, B. (1996). A review of semi-analytical solutions for flame impinge-ment heat transfer. *International Journal of Heat and Mass Transfer, 39*, 2989–3002.

Bentley, R. E. (1998). *Handbook of temperature measurement: Temperature and humidity meas-urement*. New York: Springer.

Bird, R. B., Stewart, W. E., & Lightfoot, E. N. (2006). *Transport phenomena*. New York: Wiley.

Blevins, L. G., & Pitts, W. M. (1999). Modeling of bare and aspirated thermocouples in compart-ment fires. *Fire Safety Journal, 33*, 239–259.

Boulet, P., Parent, G., Acem, Z., Collin, A., & Séro-Guillaume, O. (2011). On the emission of radiation by flames and corresponding absorption by vegetation in forest fires. *Fire Safety Journal, 46*, 21–26.

Boulet, P., Parent, G., Collin, A., Acem, Z., Porterie, B., Clerc, J. P., et al. (2009). Spectral emission of flames from laboratory-scale vegetation fires. *International Journal of Wildland Fire, 18*, 875–884.

Brohez, S., Delvosalle, C., & Marlair, G. (2004). A two-thermocouples probe for radiation cor-rections of measured temperatures in compartment fires. *Fire Safety Journal, 39*, 399–411.

Bryant, R., Womeldorf, C., Johnsson, E., & Ohlemiller, T. (2003). Radiative heat flux measure-ment uncertainty. *Fire and Materials, 27*, 209–222.

Bryant, R. A. (2009). A comparison of gas velocity measurements in a full-scale enclosure fire. *Fire Safety Journal, 44*, 793–800.

Cox, G., & Chitty, R. (1985). Some source-dependent effects of unbounded fires. *Combustion and Flame, 60*, 219–232.

Drysdale, D. (1985). *An introduction to fire dynamics*. New York:Wiley.

Dupuy, J. L. & Maréchal, J. (2010). Slope effect on laboratory fire spread: Contribution of radia-tion and convection to fuel bed pre-heating. *International Journal of Wildland Fire, 8*, 1–13.

Dupuy, J. L., Maréchal, J., & Morvan, D. (2003). Fires from a cylindrical forest fuel burner: Combustion dynamics and flame properties. *Combustion and Flame, 135*, 65–76.

Frankman, D., Webb, B. W., & Butler, B. W. (2010). Time-resolved radiation and convection heat transfer in combusting discontinuous fuel beds. *Combustion Science and Technology, 182*, 1391–1412.

Holmberg, D. G., Womeldorf, C. A., & Grosshandler, W. L. (1999). Design and uncertainty analysis of a second-generation convective heat flux calibration facility. In: *Proceedings American Society of Mechanical* Engineers (*ASME*) *Heat Transfer Division* (pp. 65–70).

Incropera, F. P., & Dewit, D. P. (2002). *Fundamentals of heat and mass transfer*. New York: Wiley.

Jou, D., Casas-Vázquez, J., & Lebon, G. (Eds.). (1996). *Extended irreversible thermodynamics*. Berlin: Springer.

Kuo, C. H., & Kulkarni, A. K. (1991). Analysis of heat flux measurement by circular foil gages in a mixed convection/radiation environment. *Journal Name: Journal of Heat Transfer (Transactions of the ASME (American Society of Mechanical Engineers), Series C); (United States); 113:4*, 1037–1040.

Lozano, J., Tachajapong, W., Weise, D. R., Mahalingam, S., & Princevac, M. (2010). Fluid dynamic structures in a fire environment observed in laboratory-scale experiments. *Combustion Science and Technology, 182*, 858–878.

Manzello, S. L., Cleary, T. G., Shields, J. R., & Yang, J. C. (2006). Ignition of mulch and grasses by firebrands in wildland–urban interface fires*. *International Journal of Wildland Fire, 15*, 427–431.

Mell, W. E., Manzello, S. L., Maranghides, A., Butry, D., & Rehm, R. G. (2010). The wildland–urban interface fire problem—current approaches and research needs. *International Journal of Wildland Fire, 19*, 238–251.

Luo, Mingchun. (1997). Effects of radiation on temperature measurement in a fire environment. *Journal of Fire Sciences, 15*, 443–461.

Morandini, F., & Silvani, X. (2010). Experimental investigation of the physical mechanisms governing the spread of wildfires. *International Journal of Wildland Fire, 19*, 570–582.

Morandini, F., Silvani, X., Rossi, L., Santoni, P.-A., Simeoni, A., Balbi, J.-H., et al. (2006). Fire spread experiment across Mediterranean shrub: Influence of wind on flame front properties. *Fire Safety Journal, 41*, 229–235.

Morandini, F., Silvani, X., Honore, D., Boutin, G., & Susset, A. (2012a). Diagnostics non intrusifs couplés des champs dynamiques et scalaires de flames d'incendie naturel en propagation. In: *13ième French Conference on Laser Measurement Techniques, Rouen*. To be edited.

Morandini, F., Silvani, X., & Susset, A. (2012b). Feasibility of particle image velocimetry in vegetative fire spread experiments. *Experiments in Fluids, 53*, 237–244.

Nathan, G. J., Kalt, P. A. M., Alwahabi, Z. T., Dally, B. B., Medwell, P. R., & Chan, Q. N. (2012). Recent advances in the measurement of strongly radiating, turbulent reacting flows. *Progress in Energy and Combustion Science, 38*, 41–61.

Parent, G., Acem, Z., Lechêne, S., & Boulet, P. (2010). Measurement of infrared radiation emitted by the flame of a vegetation fire. *International Journal of Thermal Sciences, 49*, 555–562.

Pitts, W. M. (1991). Wind effects on fires. *Progress in Energy and Combustion Science, 17*, 83–134.

Pitts, W. M., Murthy, A. V., de Ris, J. L., Filtz, J.-R., Nygård, K., Smith, D., et al. (2006). Round robin study of total heat flux gauge calibration at fire laboratories. *Fire Safety Journal, 41*, 459–475.

Santoni, P.-A., Marcelli, T., & Leoni, E. (2002). Measurement of fluctuating temperatures in a continuous flame spreading across a fuel bed using a double thermocouple probe. *Combustion and Flame, 131*, 47–58.

Schulz, C., & Sick, V. (2005). Tracer-LIF diagnostics: quantitative measurement of fuel concentration, temperature and fuel/air ratio in practical combustion systems. *Progress in Energy and Combustion Science, 31*, 75–121.

Silvani, X., & Morandini, F. (2009). Fire spread experiments in the field: Temperature and heat fluxes measurements. *Fire Safety Journal, 44*, 279–285.

Silvani, X., Morandini, F., & Dupuy, J.-L. (2012). Effects of slope on fire spread observed through video images and multiple-point thermal measurements. *Experimental Thermal and Fluid Science, 41*, 99–111.

Silvani, X., Morandini, F., & Muzy, J.-F. (2009). Wildfire spread experiments: Fluctuations in thermal measurements. *International Communications in Heat and Mass Transfer, 36*, 887–892.

Chapter 3
Beyond Measurement Devices

At the field scale, a strongly burning fire is a turbulent flow (Pitts 1991). In the strictest sense, there is no scientific definition of a turbulent flow regime, only a set of properties. For fire, these properties are described empirically through the following scenario (Santoni et al. 2006): in the gas phase, fire can extend over a large range of space and time scales that may extend up to three decades or more. These scales are organised into a cascade in which large-scale vortices transfer a large amount of mechanical energy to smaller vortices, which dissipate the energy through viscous forces. This cascade scales with a power law, in frequency or wavelength. This is a theoretical description inspired the turbulence theory of steady flows derived from the nature of local budget equations. However, few studies have evaluated empirical evidence related to this behaviour, mainly because of the lack of significant data that are available to edit the laws. Nonetheless, the turbulent nature of the fire flow can be observed visually. Signals produced by a fire are strongly variable, with more or less intense fluctuations occurring with thermal transfer (Silvani et al. 2009), even in pool fires at small scales (Consalvi 2012).

Therefore, beyond the question of how experimental data can be produced using a convenient tool, we must exercise caution in the post-processing of the results. The stochastic nature of the data must be investigated to improve our understanding of fire dynamics and to build empirical relationships that will be valuable for models.

3.1 Post-Processing Field-Scale Data: The Stochastic Nature of the Data

3.1.1 The Unsteady Nature of Fire Data

Empirical evidence of the turbulent nature of a fire spreading at the field scale has been identified. It is possible that changing the scale of fire may increase the thermal fluctuations, making its behaviour harder to predict. One can observe that

X. Silvani, *Metrology for Fire Experiments in Outdoor Conditions*,
SpringerBriefs in Fire, DOI: 10.1007/978-1-4614-7962-8_3,

the amplitudes of the fluctuations in temperature increase with the scale of the fire from roughly 200 °C up to 400 °C, around the filtered curve derived from experiments 1–4 (Fig. 2.4).

Another point relative to the stochastic nature is frequently neglected: there is no steady behaviour that lasts for the duration of experiments. For example, when collecting wind data at 1 Hz, one can assert that if there is no radical change in the wind flow (over a period of 10 min, for instance, which is the duration of a fire experiment that burned over 2000 m² of Mediterranean shrub), the data can be decomposed into an average, time-independent part and a fluctuating part, with null average. From this point of view, over a period of 10 min, the wind velocity is a statistically homogeneous flow; thus, an average naturally exists that is independent of time. This is not the case with the thermal data from a fire. Indeed, every fire quantity recorded by a sensor (mass loss, heat flux, temperature, air flow), regardless of its level of fluctuation, will evolve with decreasing distance to the fire, up to an extreme value. This means that, in a statistical sense, fire is not a steady phenomenon; there is no time-independent average. This feature must be considered at the field scale to achieve a good data-sampling rate, and post-processing must be used to account for the fluctuations.

3.1.2 Setting the Sampling Rate

The stochastic nature of fire data and its effect on the sampling rate is discussed in a pioneering work by Frankman et al. (2012). They showed that, by making the sampling rate vary during fire experiments in the open, the peak of heat radiation recorded during experiments becomes sensitive up to 5 Hz whereas the connective heat transfer significantly changes from one measurement to another at the same location if the frequency is lower than 100 Hz. This underscores absolute necessity of capturing high-frequency data regarding the hydrodynamic nature of a fire in a turbulent regime. A sampling rate of approximately 1 Hz is not sufficient for capturing important details about heat transfer. This affects the whole data-acquisition system; indeed, fire experiments in the field are about 1 h long, from the beginning of the record to the end of the fire. This means that the data logger must be able to record each differential output in a single measurement at 5 Hz. When digitised over 12 bits, this leads to about 72 MB of data. When multiplied by the number of pointwise sensors, which may easily equal 10, the amount of data will increase rapidly. Managing fire experiments requires larger memory storage, eventually with add-on memory systems. This impacts significantly the cost of the overall experiment. Another reason to set a good sampling rate for fire experiments in the field is that it allows signals to maintain regularity; i.e. there is no singularity in the form of a sudden rise in temperature, as has been observed in several experimental studies (Fig. 2.4), when the fire comes into contact with a TC. The presence of these types of stiff parts in a signal curve calls into question the use of spectral tools such as Fourier analysis, which theoretically require $C\infty$ functions. This quality forces us to avoid stiff numerical signals.

3.1.3 Investigating Fluctuations

Capturing fluctuations will be the next step for making reliable fire experiments in outdoor conditions. However, the question of their analysis remains open. In the large-scale fire experiment by Morandini et al. (2006), the use of a low-pass filter—such as the fourth-order Butterworth filter—allows for a detailed evaluation of trends in each type of turbulent data and their fluctuations. Coupling between the velocity of the incident wind and the flame temperature can be observed in these trends more easily than through a direct auto-correlation, which is hard to compute with non-stationary signals.

Because of the unsteady nature of the slow trend in fire, we expect to see the use of time-frequency signals analysis in fire data. This is not trivial because the amount of information we obtain in a wavelet transform, for instance, must be managed with caution. However, taking into account the unsteady nature of the data and assuming sufficient time and frequency resolutions, we can expect that a new type of scale will be built for interpreting signals gained in fire experiments in the open. This remains, however, a prospective point of view.

3.2 New Trends in Fire Sensing

3.2.1 Wireless Intrusive Sensors

Fire experiments in outdoor conditions are time-consuming and expensive in man-day costs because they require protecting electrical equipment from the fire. When extended surfaces of fuel are used, lengths of wires must be protected and buried to preserve the communication between sensors and data loggers. To reduce the time and labour costs of this aspect of the experiments, researchers have explored wireless solutions to fire sensing. Pioneering work began approximately five years ago, when development kits for wireless network sensors (WSN) started to be available. In (Antoine-Santoni et al. 2009), the authors protect wireless sensor nodes in a 'sock' of ceramic wood (Fig. 3.1). Displayed in

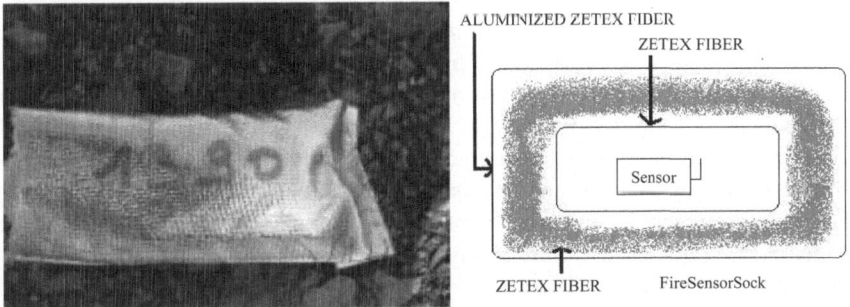

Fig. 3.1 Fire sensor sock for fire detection

Fig. 3.2 Prototype of
wireless heat sensor

outdoor conditions, the resulting four-node network allows accurate measurement
of the rate of spread (ROS) with an accuracy of about 1 % when compared to
wired sensors. In standard configurations, motes integrate sensors for air temperature and moisture content and sometimes include CO and CO_2 detectors.

WSN prototypes now exist for fire detection: the fire is 'detected' each time a sensor
breaks down. This strategy promotes spurious alerts in the case of sensor failure, and
thus requires human controls. However, there have been attempts to replace wired data
loggers with a wireless system (Fig. 3.2), for accurate and non-destructive measurements that may make it possible to turn the WSN detector into a measurement system.
In this case, the sensor—namely the measurement tool and the radio emitter—must be
protected so that they may resist the fire. Furthermore, the question of frequency sampling must be discussed carefully. Indeed, a WSN system transmits data to a collector
plugged into a PC on which a database server is running. The delay between the emission of a message containing measurement information and its reception by the PC is
the effective time interval. This means that the frequency-sampling rate depends on the
nature of the wireless communication and especially the inter-message time interval.
Standard configurations of WSN motes set an emission every Hertz. Thus, a delay of
2 s may exist between the emission and the collection of the message. The resulting
frequency rate is, therefore, maximised at 0.25 Hz according to Shannon's rule.

3.2.2 Optical Non-Intrusive Diagnostics

Recent advances have been made in fire sciences concerning the measurement of
the velocity vector field using two or three components. The use of PIV allows for
the production of tomographic images of flames containing luminous filaments of
soot, from which the mixing structure of the flow is revealed. Because of its porta-
bility to intermediate-scale experiments (i.e. vegetal plots of about 100 m²; Table 1),
the PIV can go from lab to outdoor conditions. A remaining problem concerns the
ability of PIV cameras to capture a truncated range of spatial scales, because of limitations in CMOS quality. The next generation of tools will provide velocity field
measurements with a resolution of about 30 Mp at length scales up to two decades.

As noted previously (see 'Measuring scalars'), the measurement of the gas-mixture fraction, soot-volume fraction, and the concentration of given species is not trivial. If modern diagnostics exist for strongly radiative turbulent flames (Schulz and Sick 2005), the extrapolation of these diagnostics to large-scale fire scenarios is a work-in-progress. One can assume that future progress in optical light sources and computing power will lead to cost reductions that will make such adaptations easier in the coming years.

Along with measurements of the velocity fields, observing the mixing process in a combustion system in a real-scale fire may be the most important diagnostic tool in the field of turbulent combustion. Although the mechanical processes through which turbulence mixes are not well known, it is of primary interest to examine the internal structure of a flame, to understand how reactive gases burn. Optical diagnostics based on laser lighting and dedicated digital cameras (advanced or not) are excellent developments for non-academic cases. Other techniques exist without light sources. For instance, the background-oriented schlieren technique captures the refractive index of a fluid modifying the speckle aspect observed by the digital camera. The corresponding field can be related to the density field of the flow. Examples in which such techniques have been used in fire analyses are rare. One can cite (Deimling et al. 2011) for the case of a hydrogen fire jet and readers can find a recent overview of quantitative measurements with schlieren techniques in (Hargather and Settles 2012).

3.2.3   Remote Sensing

Fire results from the influence of many parameters; among these, the upwind wind flow must be investigated in experimental campaigns at the same time as thermal quantities and fire kinematics. Wind anemometers may be installed around the experimental plot. Sonic anemometers are devices that measure the velocity using sound waves; they have a very fine temporal resolution, greater than 20 Hz, yielding turbulence features.

When fires propagate at the scale of a valley, experimental studies may use meteorology instruments. Sonic detection and ranging (Sodar) works like radar but uses sound waves rather than radio waves. It allows the measurement of the wind profile in the surface atmospheric boundary layer. This information is of primary importance for modelling fire spread at such a scale.

In the same category is light detection and ranging (Lidar), a remote-sensing system based on the reflection of laser light by a distant object. With regard to fire experiments in the field, it can be used to ascertain further the wind profile; in this case, the velocity measurement is based on the Doppler effect on water molecules or small particle transported by the wind. Based on the theory of relativity, there is a linear dependence between the velocity of an object (if that velocity is small compared to the light velocity, c) and the frequency of the light emitted by this same object. Measuring the frequency shift to the source (with an interferometer, for instance) leads to an accurate measurement of the velocity of the object. Lidar also can be used to track particulates in the fire plume to measure size and concentration; in this case, it is called aerosol Lidar.

Lidar, therefore, can detect the complex vertical structure of the atmosphere with high spatial and temporal resolution, making it well suited for understanding atmospheric dynamics and transport processes. However, it has two main disadvantages: it is expensive and the signals produced must be interpreted with caution. A single configuration with direct smoke detection is under development for a ground-based configuration. For the technique to become better known, many technical problems need be addressed, especially eye protection from pulse light sources, as discussed in (Utkin et al. 2003). Studies of the structure of smoke plumes can also be performed using direct smoke detectors (Lavrov et al. 2006).

3.3 Managing the Fire Experiments

Ultimately, despite the difficulties of fire experiments in the field, intense effort has been made in measurement, post-processing, and the communication of physical quantities. However, this work is not sufficient to ensure the success of fire experiments in the open. The appropriate management of such events is also a subject for discussion.

3.3.1 Meteorological Conditions

Fire experiments in the open depend on meteorological conditions, which have a direct effect on the experimental strategy. We can report our own experience with fire experiments in the northern Mediterranean region that took place from 2003 to 2007. Figure 3.3 illustrates the impact of the metrology strategy on the number, cost (in men per event), and the size of the fire experiment. We

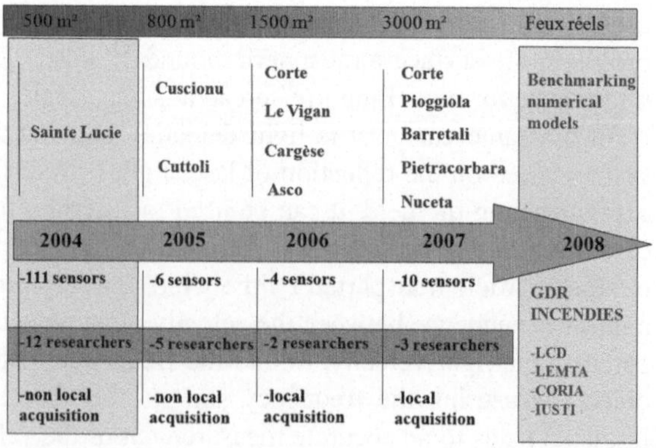

Fig. 3.3 Evolution of the large-scale experiments performed in the SPE Lab starting in 2004

500 m²	800 m²	1500 m²	3000 m²	Feux réels
Sainte Lucie	Cuscionu Cuttoli	Corte Le Vigan Cargèse Asco	Corte Pioggiola Barretali Pietracorbara Nuceta	Benchmarking numerical models
2004	**2005**	**2006**	**2007**	**2008**
-111 sensors	-6 sensors	-4 sensors	-10 sensors	GDR INCENDIES
-12 researchers	-5 researchers	-2 researchers	-3 researchers	-LCD -LEMTA -CORIA -IUSTI
-non local acquisition	-non local acquisition	-local acquisition	-local acquisition	

started in 2005 with a relatively small experiment on a parcel located 3 h away, by car, from our lab, which employed our whole team over a period of four months. Because of permanent contradictory wind conditions (the predominant wind direction was downward), ignition was frequently reported. Setting up 111 sensors and protecting about 500 m of wires accounted for the majority of our time on the ground. In 2007, using a new sensor with no additional wire and focusing on 10 measurement points, we performed five experiments on larger areas with just three individuals. An empirical rule for reducing the requirements for team and tools could be 'installing and setting the whole metrology up in one hour before the first ignition'. In 2012, using this strategy, we have shared experiments in Corsica with colleagues from fire-science laboratories in France at the the Centre National de la Recherche Scientifique, i.e. LCD, LEMTA, CORIA, IUSTI. We can drive them to an experimental fire plot that is located a distance of 1 h from local airports.

In order to adapt the experiment f to the variability of meteorological conditions, the design and set-up of a fire experiment (i.e. the team size and measurement tools) must be initiated 15 days before the experiments, controlled every day (especially with regard to the battery load and clock synchronisation of all electronic equipment) and, optimally, each device must be ready 3 days prior to the experiment. In these conditions, one minimizes the risks of cancellation because of failure, the time for installing/uninstalling the devices, and the loss of data. The length of this period must be clearly defined and cannot exceed 10 days after the initial date because of the psychological pressure it puts on the entire team over a long period.

3.3.2 Cost

Cost is a central question for large-scale fire experiments. One large-scale scenario, the international Crown Fire Modeling Experiment [ICFME; (Butler et al. 2004)] is rumoured to cost USD 1 million! Such a large scale experiment is hard to duplicate. Compressing the time to set up the experiment, according to the strategy described above, has a global positive impact on the overall cost. This reduces the number of individuals involved and the cost of the metrology, even if, for instance, the abandonment of wired systems also impacts the frequency-resolution quality of the data obtained.

Because of these costs, we plan to develop mobile systems that are capable of studying real-scale fires; firemen and their fire-fighting systems (such as ground and air vehicles) provide a natural opportunity to sense fires under real-world conditions. The goal of the REC1/MICNA project (Fig. 3.4), which began in 2011, is to develop an all-terrain vehicle for insertion into large-scale fires. It is fire-resistant and it has been turned into a mobile laboratory for heat and chemical measurements. Such a solution should help us increase the collection of empirical data from fires in real-scale conditions.

Fig. 3.4 REC1/MICNA
prototype

Cost reductions should also be supported by the manufacturers of fire-fighting systems and fire-fighter protection. Measuring physical quantities at the field scale will help industrial partners to upgrade the efficiency of their equipment by providing valuable data. Because fire science is an engineering science, a closer relationship should be built between industry and the scientific laboratories that study fires in the field.

References

Antoine-Santoni, T., Santucci, J.-F., de Gentili, E., Silvani, X., & Morandini, F. (2009). Performance of a protected wireless sensor network in a fire. Analysis of fire spread and data transmission. *Sensors, 9*(8), 5878–5893.

Butler, B. W., Cohen, J., Latham, D. J., Schuette, R. D., Sopko, P., Shannon, K. S., et al. (2004). Measurements of radiant emissive power and temperatures in crown fires. *Canadian Journal of Forest Research, 34*, 1577–1587.

Consalvi, J. L. (2012). Influence of turbulence–radiation interactions in laboratory-scale methane pool fires. *International Journal of Thermal Sciences, 60*, 122–130.

Deimling, L., Weiser, V., Blanc, A., Eisenreich, N., Billeb, G., & Kessler, A. (2011). Visualisation of jet fires from hydrogen release. *International Journal of Hydrogen Energy, 36*, 2360–2366.

Frankman, D., Webb, B. W., Butler, B. W., Jimenez, D., & Harrington, M. (2012). The effect of sampling rate on interpretation of the temporal characteristics of radiative and convective heating in wildland flames. *International Journal of Wildland Fire*, in press.

Hargather, M. J., & Settles, G. S. (2012). A comparison of three quantitative schlieren techniques. *Optics and Lasers in Engineering, 50*, 8–17.

Lavrov, A., Utkin, A. B., Vilar, R., & Fernandes, A. (2006). Evaluation of smoke dispersion from forest fire plumes using lidar experiments and modelling. *International Journal of Thermal Sciences, 45*, 848–859.

Morandini, F., Silvani, X., Rossi, L., Santoni, P.-A., Simeoni, A., Balbi, J.-H., et al. (2006). Fire spread experiment across Mediterranean shrub: Influence of wind on flame front properties. *Fire Safety Journal, 41*, 229–235.

Pitts, W. M. (1991). Wind effects on fires. *Progress in Energy and Combustion Science, 17*, 83–134.

Santoni, P. A., Simeoni, A., Rossi, J. L., Bosseur, F., Morandini, F., Silvani, X., et al. (2006). Instrumentation of wildland fire: Characterisation of a fire spreading through a Mediterranean shrub. *Fire Safety Journal, 41*, 171–184.

Schulz, C., & Sick, V. (2005). Tracer-LIF diagnostics: Quantitative measurement of fuel concentration, temperature and fuel/air ratio in practical combustion systems. *Progress in Energy and Combustion Science, 31*, 75–121.

Silvani, X., Morandini, F., & Muzy, J.-F. (2009). Wildfire spread experiments: Fluctuations in thermal measurements. *International Communications in Heat and Mass Transfer, 36*, 887–892.

Utkin, A. B., Fernandes, A., Simo″es, ES, F., Lavrov, A., & Vilar, R. (2003). Feasibility of forest-fire smoke detection using lidar. *International Journal of Wildland Fire, 12*, 159–166.

Chapter 4
Conclusion

Although most predictive studies of fire are performed at regional scales (Liu et al. 2010), global climate change due to greenhouse effects is expected to promote an increase in wildfire activity all over the world: European Regional Development Fund encourages international projects that develop help-to-decision tools for managing crisis, among which wildfire has an important place. Consequently, fire experiments must be performed at the field scale, to generate data that will fill the gap between what is known about the burning of solid or liquid fuels in the laboratory and these events at real scales. In the present brief, we attempt to give an overview of existing and future systems that are able to sense fires in real-scale conditions.

One can summarize this effort presenting Table 4.1 An important question is what measurements are relevant and which related devices are useful to make a given experimental fire scenario valuable. This non-exhaustive overview treats only a few subjects from the laboratory scale to class 3 fire experiments (1000–10000 m^2). There is no strategy for directly sensing fire at real scales, during a real fire scenario. This is mainly due to the fact that, by changing the scale from class 1 to 4, the fire science also changes: at the small scale, fire is related to mechanical engineering, combustion, the chemistry of reactive mixing of burning gases, and fluid mechanics. At real scales, fire interacts with the atmosphere, the topography, and the vegetal cover at the mesoscale, and thus becomes related to sciences of the universe and ecology. Experimental tools respect this gap, with each dedicated to the respective scale of each discipline.

In mechanical engineering, one can expect great advances in optical diagnostics for large-scale radiative flames. In the framework of ecology, experimental fire science must be attentive to the development of remote-sensing technologies based on Lidar.

Fire science is therefore at the crossroads of multiple types of physics and multiple scales. Despite obvious limitations in funding and support, the organisation of fire experiments should not be limited to a single fire class. On the contrary, we think that every effort in developing or promoting a new experimental technique

X. Silvani, *Metrology for Fire Experiments in Outdoor Conditions*,
SpringerBriefs in Fire, DOI: 10.1007/978-1-4614-7962-8_4,
© The Author(s) 2013

Table 4.1 Classes of fire experiments. With regard to measurements and devices, bold items are those for which results are frequently available in the fire literature

Fuel area	Indoor/outdoor	Measurement	Devices
<10 m²	Indoor Repeatable	Mass loss rate HRR/gas analysis	TGA—Cone calorimeter—Fire propagation apparatus
(10–1000 m²)	Indoor/outdoor Repeatable	Heat flux Soot production Mass loss rate Gas composition and mixing Wind profile and plume dynamics	TC, HFM, PIV, LIF Optical diagnostics Load cells Spectrometer FTIR/CEM, OH* chemiluminescence Sonic/hot wire anemometry
(1000–10000 m²)	Outdoor Non-repeatable	Heat Flux Soot production Mass loss rate Gas composition and mixing Wind profile and plume dynamics Aerosol	TC, HFM Optical diagnostics Load cells Spectrometer FTIR/CEM, OH* chemiluminescence Sonic/hot wire anemometry, Sodar, Doppler lidar Aerosol lidar
>1 ha	Outdoor Non-repeatable	Heat Flux Gas composition and mixing Wind profile and plume dynamics Aerosol	TC, HFM Spectrometer FTIR/CEM, OH* chemiluminescence Sonic/hot wire anemometry, Sodar, Doppler lidar Aerosol lidar

must keep in mind how the technique will translate to the fire class immediately above, thus considering the scale dependency. Even if it is common to pretend that fire at the real scale does not need details, we think that details may be more important than trends, because of the turbulent features of fire. Therefore, these details must be observed attentively.

Finally, one can expect that the multi-disciplinary range of the subject will lead researchers to gain data during real fires. This will require scientists to work more closely with fire-fighting and forest-management staffs. This may also produce unexpected and valuable strategies, combining the needs of science and the requirements of fire-fighting policies.

Acknowledgments The author is extremely grateful to his colleagues, Dr. Frédéric Morandini and Antoine Pieri, with whom the adventure of performing fire experiments began 10 years ago. This brief is dedicated to them. I also thank Pr. Pascal Boulet, Dr. Jean-Luc Dupuy, and Dr. Olivier Sero-Guillaume for their fruitful discussions. I also thank Arnaud Susset from the R&D vision company for helping me to understand and use optical diagnostics for reactive flows. Last but not least, a special thank you goes to Dr. Jean-François Muzy for present and future work in data processing of singular signals and the beautiful physics to which it is related.

Reference

Liu, Y., Stanturf, J., & Goodrick, S. (2010). Trends in global wildfire potential in a changing climate. *Forest Ecology and Management, 259*, 685–697.